Awesome, Disgusting Careers
DISGUSTING
ANIMAL CARE
JOBS
Stephanie Bearce
BLACK
RABBIT
BOOKS

Hi Jinx is published by Black Rabbit Books
P.O. Box 227, Mankato, Minnesota, 56002.
www.blackrabbitbooks.com

Marysa Storm, editor; Michael Sellner, designer
and photo researcher

Library of Congress Cataloging-in-Publication Data
Names: Bearce, Stephanie, author.
Title: Disgusting animal care jobs / by Stephanie Bearce.
Description: Mankato, Minnesota : Black Rabbit Books, [2023] |
Series: Hi jinx. Awesome, disgusting careers | Includes bibliographical
references and index. | Audience: Ages: 8-12 | Audience: Grades: 4-6 |
Summary: "Let readers explore the awesome, disgusting animal care jobs
that keep their world running smoothly through witty, conversational
text, fun facts, and critical thinking questions that'll have them laughing
and learning"- Provided by publisher.
Identifiers: LCCN 2020030399 (print) | LCCN 2020030400 (ebook) |
ISBN 9781623106805 (hardcover) | ISBN 9781644665473 (paperback) |
ISBN 9781623106867 (ebook)
Subjects: LCSH: Animal specialists-Vocational guidance-
Juvenile literature.
Classification: LCC SF80 .B42 2022 (print) | LCC SF80 (ebook) |
DDC 636.0023-dc23
LC record available at https://lccn.loc.gov/2020030399
LC ebook record available at https://lccn.loc.gov/2020030400

Image Credits

Alamy: Ockra, 23; Wayne HUTCHINSON, 4; Nature Picture Library, 8–9; Shutterstock: Alica in Wonderland, Cover, 12, 21; Ali Kiraz, 10; Angeliki Vel, 20; AnggaR3ind, Cover, 4, 5; Arcady, 11; Cartone Animato, 15; Chelle129, 16; Denis Zhengal, 12; ekler, 13; Elena11, 16; Elnur, 18; Fat Jackey, 15; Holger Kirk, 3; lady-luck, 14, 15, 23; LoopAll, Cover, 1, 5, 6, 20; Lorelyn Medina, 11; mconnel, 12; Memo Angeles, Cover, 5, 7, 20, 23; Mumemories, 16; Pasko Maksim, 6, 23, 24; Passakorn Umpornmaha, 12; Pattama Rattanawan, 19; Pitju, 21; Petra's Vectors, 18; ridjam, 15; Ron Dale, 5, 6, 20; Sararoom Design, Cover, 4; SeDmi, 10–11; Sukpaiboonwat, 7; Tony Oshlick, Cover, 4, 5, 14, 15, 21; totallypic, 10, 16; Witchanee.photo, 16; Zacchio, 7

CONTENTS

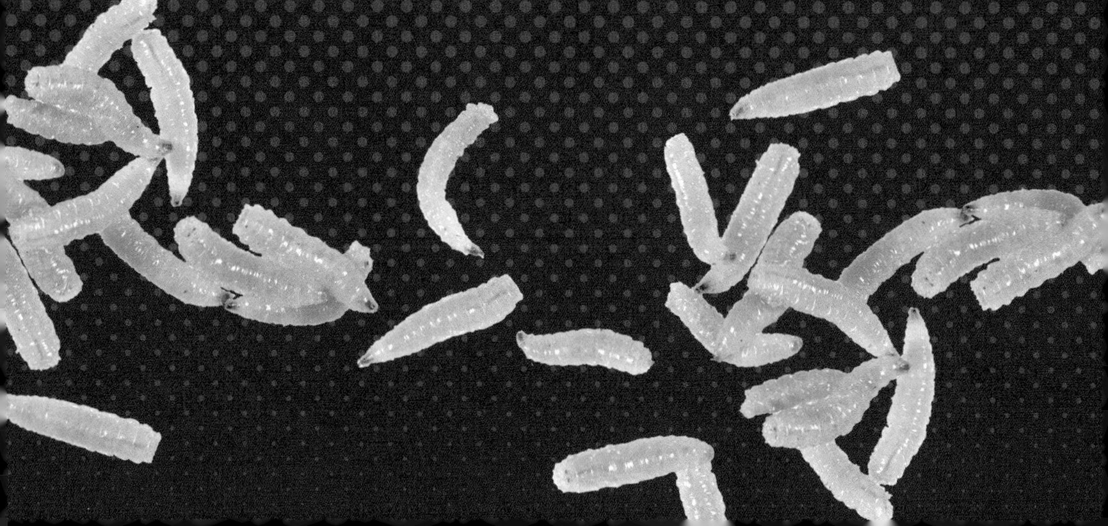

Chapter 1
WACKY WORK WITH ANIMALS

Working with animals is not all fluffy bunnies and cuddly kittens. Some animal jobs involve working with poop and pee. Others require handling poisonous snakes or blood-sucking leeches. Ready to learn more? Hold on to your stomach. Animal careers can be awesomely disgusting.

Milking Cows

Some dairy farmers use machines to milk their cows. Others milk them by hand. No matter how farmers milk, they get dirty. Milk comes from cows' udders. That means farmers must reach up by cows' butts. They must clean the **teats** and udders. Farmers might get kicked or bitten. There's also a chance cows might poop on them. For these people, it's all part of a day's work.

One dairy cow produces about 80 pounds (36 kilograms) of **manure** a day.

milking machine

milking by hand

Gaboon vipers have the longest fangs of any **venomous** snake. Their fangs reach up to 2 inches (5 centimeters) long.

Milking Snakes

If milking cows isn't extreme enough, try snake milking. Scientists and researchers milk venomous snakes for their poison. Brave workers cover the mouths of jars with plastic. Then, they grab the snakes behind the heads. Next, the milkers press the snakes' fangs into the plastic. The snakes then spit venom into the jars. The venom is often used to make **antivenom**.

Leech Farmers

Not all farmers have cows in fields. Some raise leeches in labs. Farmers **breed** leeches and wait for babies to hatch. When the babies hatch, they're hungry. Farmers must feed them animal blood. They also need to keep the tanks clean. The farmers must be careful during all this. Leeches are escape artists. They can attach to farmers' skin and hitch rides out.

leech cocoons

Leech farmers sell the little bloodsuckers to doctors. These doctors use them during some surgeries.

rotting food

bait

maggots

Maggot Farmers

The only thing grosser than raising leeches might be raising maggots. To farm maggots, workers put piles of rotting food in bins. Flies lay their eggs in this goop. The eggs then hatch into white wriggly maggots. When the maggots are big enough, the farmers collect them. They pack the little worms up to sell as animal food or bait for fishing.

Testing Animal Poop

All animals poop. And it's up to scientists and vets to test it! Testing poop can show if animals are healthy. It can tell owners if females are going to have babies. To test poop, workers must first collect it. They watch and wait for the animals to go. That way, they know which piles belong to which animals. As soon as the animals are done pooping, the workers collect the steamy samples. Then they test the poo.

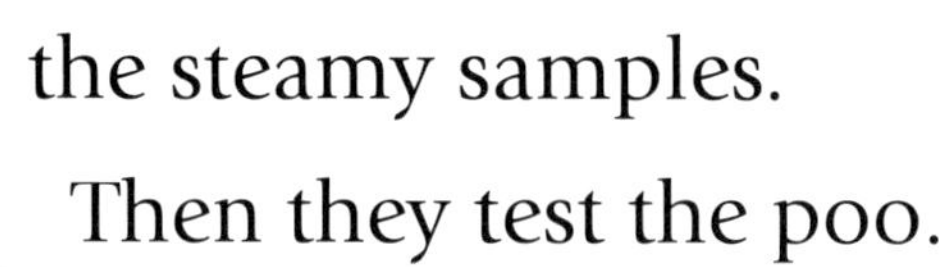

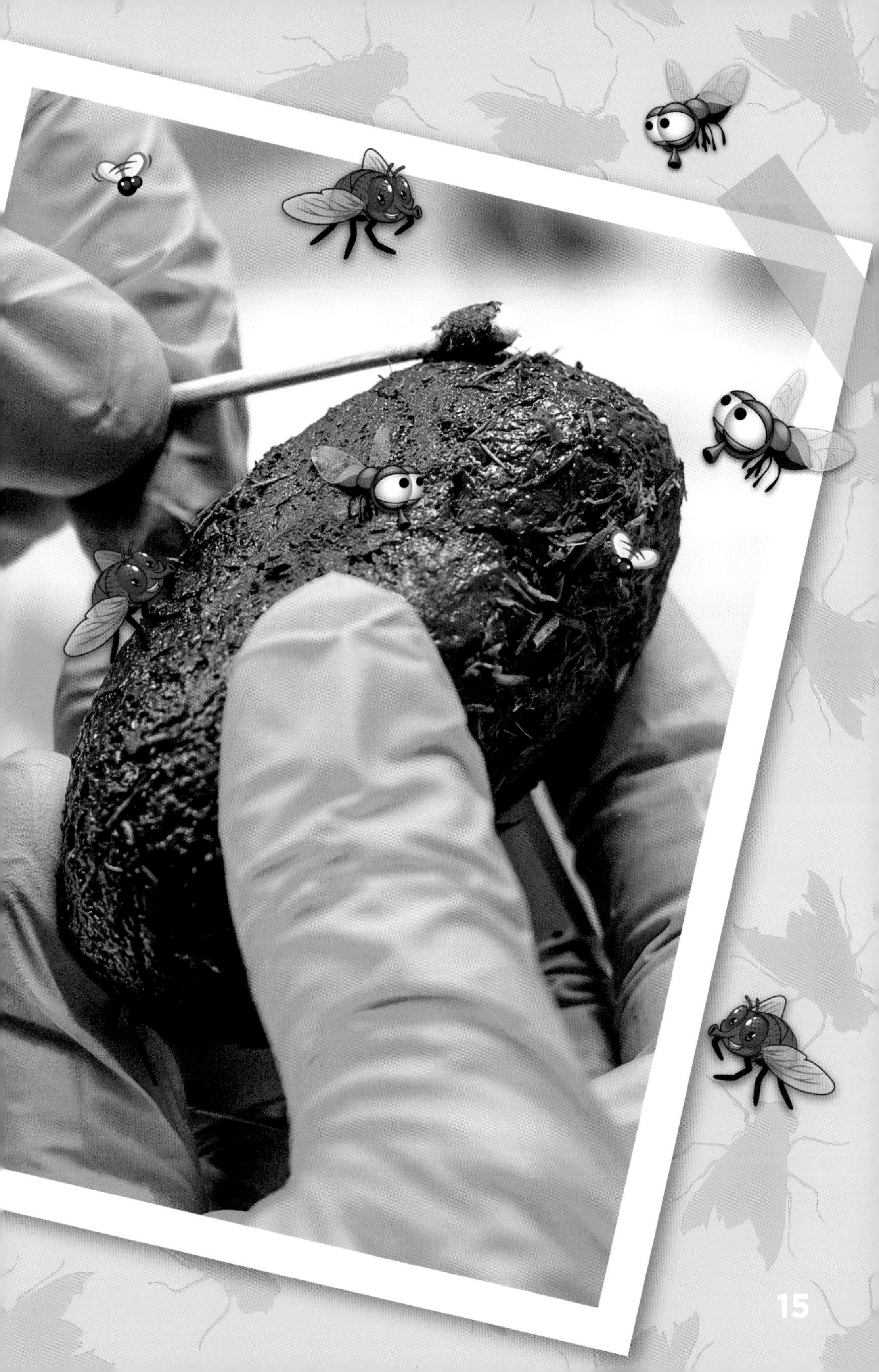

Manure
Manure Mixture
Manure Dirt
Compost

Compost Farmers

Horses are beautiful animals. But raising horses isn't a glamorous job. Horse owners must deal with a lot of manure. Some turn it into **compost**. Farmers start by gathering huge piles of the manure. The farmers add leaves and animal bedding, such as wood shavings. Then they cover the piles with giant tarps. Every few days, they use tractors to turn the piles over. Rotating the piles helps them break down evenly. Eventually, the poop starts looking like dirt. Farmers dry it out and sell it. Gardeners buy it to help their plants grow.

Urine Farmers

Don't like poop? You could always work with pee. Some people work on wild game farms. They make sure the wild animals living there are healthy. To do so, they test pee. To collect pee, the workers trap animals in pens. These traps are safe for the animals. They also have floors with drains. The animals' pee then drips through the drains and gathers in pans. Once the animals have peed, the workers release them. Then they test away!

Some game farmers bottle and sell the animal pee. Gardeners use it to keep **pests** away.

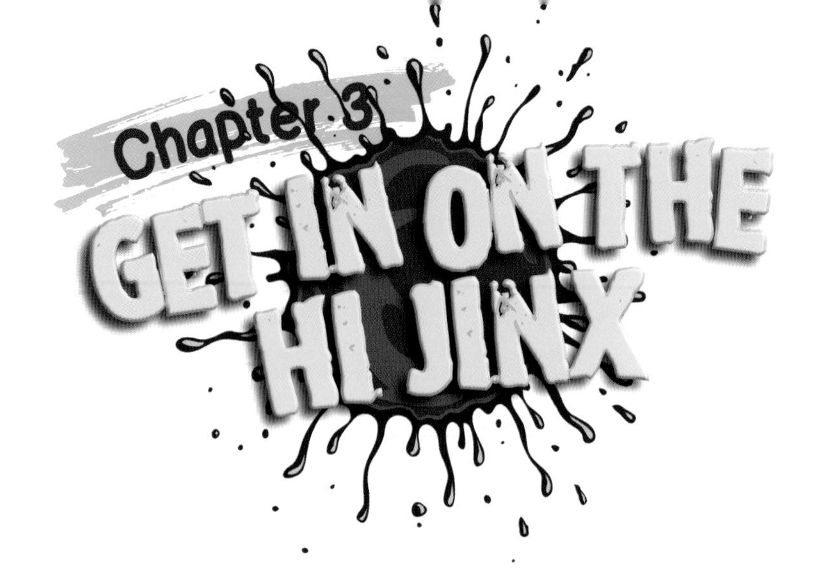

There are many interesting animal jobs. You can research them by going to your library. You can also talk to farmers, vets, and other people in your community. They can tell you all about their jobs and the training they needed. Who knows? You might find the perfect animal job for you.

Take It One Step More

1. Why is manure so good for plants? Do some research to find out.

2. Is there a job in this book that you would never do? Why?

3. Would you rather milk snakes or cows? Explain your choice.

GLOSSARY

antivenom (an-tee-VEN-uhm)—a liquid used to counteract poison from an animal, such as a snake

breed (BREED)—the process by which young animals are produced by their parents

compost (KOM-pohst)—decayed organic material, such as leaves and grass, used to improve soil

manure (muh-NOOR)—solid waste from farm animals that is used to make soil better for growing plants

pest (PEST)—an animal or insect that causes problems for people especially by damaging crops

teat (TEET)—the part of a female animal through which a young animal receives milk

venomous (VEN-uh-mus)—containing venom or poison

LEARN MORE

BOOKS

Krajnik, Elizabeth. *Dairy Farmers.* Getting the Job Done. New York: PowerKids Press, 2020.

Nickel, Scott. *World's Grossest Jobs.* Minneapolis: Lerner Publications, 2021.

Wild, Gabby, and Jennifer Szymanski. *Wild Vet Adventures.* Washington, DC: National Geographic Kids, 2020.

WEBSITES

Dairy Farming Facts for Kids
kids.kiddle.co/Dairy_farming

So You Want to Be a Zookeeper
www.stlzoo.org/animals/soyouwanttobeazookeeper

Top 10 Animal Gross-Outs
www.animalplanet.com/wild-animals/10-animal-gross-outs/

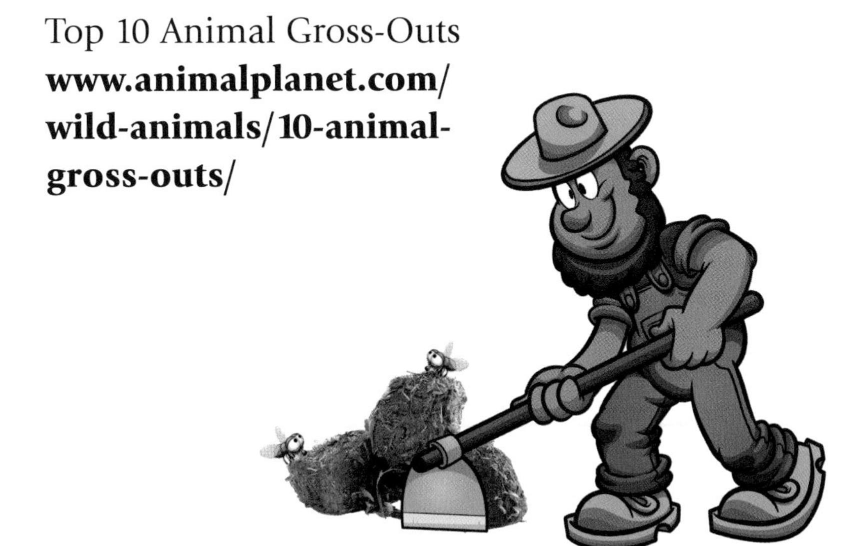

INDEX